iMath
Readers

Pools to Ponds:
Area, Perimeter, and Capacity

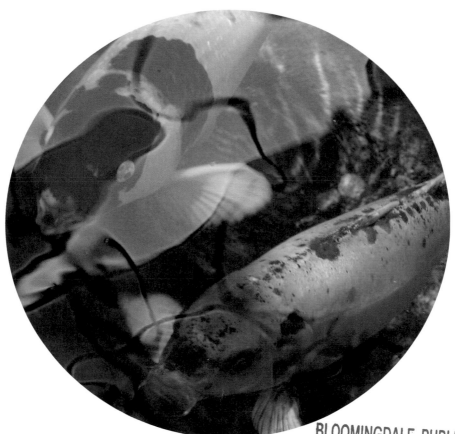

by John Perritano

Content Consultant
David T. Hughes
Mathematics Curriculum Specialist

NORWOOD HOUSE PRESS
Chicago, IL

Norwood House Press
PO Box 316598
Chicago, IL 60631

For information regarding Norwood House Press, please visit our website at
www.norwoodhousepress.com or call 866-565-2900.

Special thanks to: Heidi Doyle
Production Management: Six Red Marbles
Editors: Linda Bullock and Kendra Muntz
Printed in Heshan City, Guangdong, China. 208N—012013

Library of Congress Cataloging–in–Publication Data

Perritano, John.

Pools to ponds: area, perimeter, and capacity / by John Perritano;
content consultant, David Hughes, mathematics curriculum specialist.
pages cm.—(imath)

Summary: "The mathematical concepts of perimeter, area, and volume are
introduced as a child constructs ponds for fish and ducks. Readers learn the
formulas for these measurements, while also learning about the history of
pools and swimming. Includes a discover activity, history connection, and
mathematical vocabulary introduction"—Provided by publisher.

Includes bibliographical references and index.

ISBN: 978-1-59953-560-9 (library edition: alk. paper)
ISBN: 978-1-60357-529-4 (ebook)

1. Area measurement—Juvenile literature.
2. Volume (Cubic content)—Juvenile literature.
3. Volumetric analysis—Juvenile literature. I. Title.

QC104.5.P47 2013
530.8'1—dc23
2012025623

CONTENTS

Note to Caregivers:

Throughout this book, many questions are posed to the reader. Some are open-ended and ask what the reader thinks. Discuss these questions with your child and guide him or her in thinking through the possible answers and outcomes. There are also questions posed which have a specific answer. Encourage your child to read through the text to determine the correct answer. Most importantly, encourage answers grounded in reality while also allowing imaginations to soar. Information to help support you as you share the book with your child is provided in the back in the **Additional Notes** section.

Bold words are defined in the glossary in the back of the book.

Summer's Here

Yippee! Summer is almost here. The sun is bright. The air is hot. Let's get in the pool!

Ted and Patty own their own pool business. They design *and* build pools and ponds. They have the nicest pool. They play music at their pool parties. They listen to a band called the Beach Boys. Their nephew James has never heard of that band before. Uncle Ted says the group is "groovy." James doesn't know what that means exactly, but it sounds fun.

Uncle Ted and Aunt Patty usually build huge pools. But James wants them to help him build something smaller.

Ponds, actually. One for his fish. One for his ducks. And a small pool for him.

Boxed In

Think of a pool as a big box. It has **length**, **width**, and **depth**. The distance around the box, or pool, is the pool's **perimeter**. How much space a pool covers is its **area**. And how much water it can hold is its **liquid volume**, or **capacity**.

You can use different units to measure capacity.

Here are some **customary units of capacity**.

teaspoon (tsp)	tablespoon (Tbsp) 1 Tbsp = 3 tsp	cup (c) 1 c = 16 Tbsp
pint (pt) 1 pt = 2 c	quart (qt) 1 qt = 2 pt	gallon (g) 1 g = 4 qt

Here are some **metric units of capacity**.

liter (L)
milliliter (mL) 1 mL = 1/1000 of a liter

How do builders use perimeter, area, and capacity to build a pond or pool? Let's find out.

Perimeter: Builders can measure and add the lengths of the sides of a pool to find its perimeter.

The formula $l + l + w + w$ **or** $2l + 2w$ may be used to find the perimeter of a rectangle. The l stands for length, and the w stands for width.

Which way would you use to find the perimeter of a pool?

l

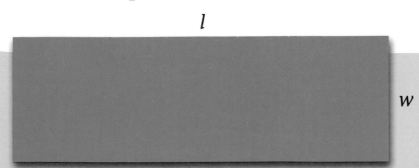

w

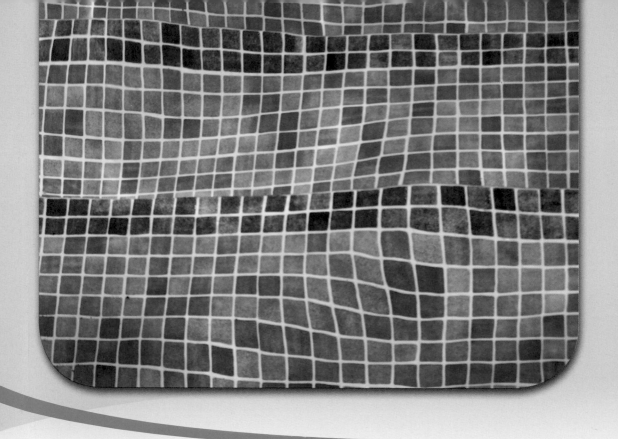

Area: Builders can cover a pool's surface with unit squares to find its area. Then, they can count the squares. The sum is the pool's area. It is expressed in square units.

Or: Builders can use the formula $l \times w$. The l stands for length, and the w stands for width.

Which way would you use to find the area of a pool?

8

Liquid Volume, or Capacity: Builders can use customary units of capacity to measure how much water a pool can hold.

Or: Builders can use metric units of capacity to measure how much water a pool can hold.

Which units would you use to measure capacity?

DISCOVER ACTIVITY

Measure by Hand

Find three plastic rectangular boxes in your house. Look for the kinds of boxes you use to store leftover food. Plastic tubs will work, too.

Use a ruler to measure the length, width, and height of each box. Record the measurements. Remember to include the units you used.

Materials

- 3 plastic rectangular boxes
- ruler marked in either customary or metric units
- liquid measuring tools marked in either customary or metric units
- water
- paper towels

Use a measuring cup to determine the liquid volume, or capacity, of each box. Record the measurements. Remember to include the units you used.

Use the paper towels to clean up any water spills.

How will you determine the perimeter, area, and liquid volume, or capacity, of each box? Will you:

- use one of these formulas to find the perimeter?
    $l + l + w + w$
    or
    $2l + 2w$

- use one of these methods to find the area?

    count square units
    or
    use the formula $l \times w$

How will you determine the units of capacity for each box? Will you:

- use customary units of capacity?

- use metric units of capacity?

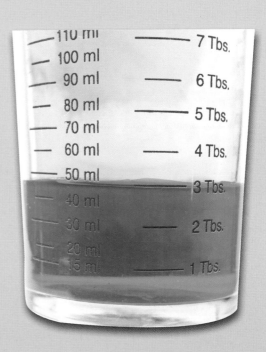

"Fintastic" Fish

James loves fish. His favorite is Myrtle, his goldfish. She sits in a bowl on his dresser.

James wonders if Myrtle would like a tank instead of a bowl. He finds a tank 15 inches wide and 10 inches long. What is the tank's perimeter? Use the formula $l + l + w + w$ to find the answer. Then, use the formula $2l + 2w$. Do the answers match?

James also likes koi. They are colorful fish from Japan. He hopes Uncle Ted and Aunt Patty can design a pond for the koi.

Koi do well in large ponds. They can grow up to 2 feet in length. Myrtle is the size of James's thumb.

Uncle Ted and Aunt Patty plan a way to keep fresh water flowing through a koi pond. They also draw a plan for the koi pond's shape and size. What is the pond's perimeter? What is its area?

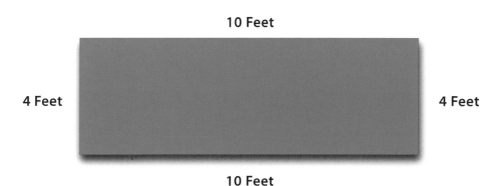

10 Feet

4 Feet

4 Feet

10 Feet

"Deep-see" Fishing

The pond has to be deep enough for the fish to swim. So, Aunt Patty and Uncle Ted make it 3 feet deep.

James thinks about which customary unit of capacity he could use to measure how much water the pond could hold. Which unit holds the most water?

teaspoon (tsp)	tablespoon (Tbsp) 1 Tbsp = 3 tsp	cup (c) 1 c = 16 Tbsp
pint (pt) 1 pt = 2 c	quart (qt) 1 qt = 2 pt	gallon (g) 1 g = 4 qt

 Did You Know?

Koi are as important to their Japanese families as dogs and cats are to their U.S. families. But koi often live much longer than their owners. Researchers took scales from koi living in a pond in a Japanese village. The oldest koi was 226 years old! Two other fish in the pond were 180 and 156 years old. So, people pass their fish on to their children. They, in turn, pass the fish to their children, and so on.

Just Ducky!

Koi are cool, but ducks are awesome! James likes ducks, too!

Mallard ducks fly around his house all the time, and he would like to give them a nice place to land. Then, maybe they'll stay to eat. Or maybe they'll make nests.

A male mallard has a bright green head and a yellow bill. He has a chestnut-colored chest, a gray body, and a black rear.

A female mallard is brown with an orange and brown bill. Like the male, the female has a blue patch on each wing. The patch is bordered in white.

Mallards can live almost anywhere, even in someone's backyard.

Mallards mate for life. They make their nests on the ground. When they breed, they lay about a dozen eggs. It takes about a month for the eggs to hatch. The plan is to plant tall grasses around the pond so the ducks can hide their nests.

Mallards are "dabbling ducks." That means that they tip forward into water to eat underwater plants.

The duck pond that Uncle Ted and Aunt Patty are building needs to have a wide area. That's because mallards also paddle on the surface to find food.

Uncle Ted and Aunt Patty draw two designs for the duck pond. What is the area of each pond? Use square units to find the area of the first design. Use the formula $l \times w$ to find the area of the second design.

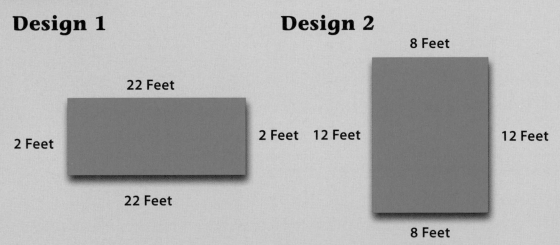

Design 1

22 Feet

2 Feet

22 Feet

Design 2

8 Feet

2 Feet   12 Feet

12 Feet

8 Feet

They decide to build the second pond. It has more room for the ducks to splash.

James is very excited about the design Uncle Ted and Aunt Patty chose for the duck pond. The pond gives the ducks plenty of room and plenty of water.

James plants grass around the pond. Maybe the ducks will build their nests there.

He wants to keep the nests safe from feet, like his brother's. He wants to protect them from hungry animals, too. So, James designs a fence around the pond's perimeter.

The pond is 12 feet long by 8 feet wide. James needs 3 more feet from the edge of the pool to the fence line. Then, he can build the fence. How much fencing does he need to go all around the pond's perimeter?

A Pool for James

Why should the koi and mallards have all the fun? James wants his Uncle Ted and Aunt Patty to design and build a pool for him, too!

They look at some plans together. "How wide do you want your pool to be, James?" they ask. "How long should we make it? Do you want a wide area for playing water games? Or do you want a narrow one for swimming laps? We need to figure out the area of your pool before we begin digging."

Uncle Ted adds, "How much water do you want the pool to hold? Do you want to keep it shallow for games? Or do you want it to be deep for diving?"

Uncle Ted's and Aunt Patty's questions give James something to think about. How big *does* he want his pool to be? Before deciding, James does some reading about pools, including really old ones.

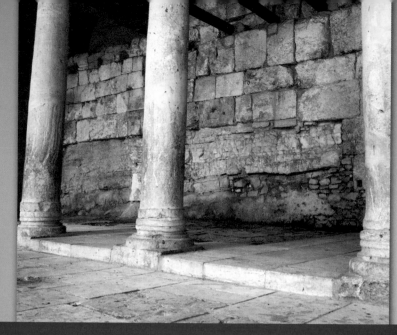

Some of the Roman soldiers of the Tenth Legion may have walked down this old main market road through Jerusalem.

Connecting to History

Jerusalem is an ancient city in Israel. Builders there planned to build a bath. But scientists found an ancient swimming pool beneath the ground. Roman soldiers built it about 2,000 years ago. The pool has a **mosaic** floor and clay tiles. The tiles are stamped with the name of the soldiers' unit. The soldiers belonged to the Tenth **Legion**.

The pool was used for more than swimming. Soldiers used it for bathing, too.

The builders will build the bath when the scientists are finished with their work. But the builders will combine parts of the old bath with the new.

Diving into History

There is an ancient city in Pakistan. This city is called Mohenjo-daro. The words mean "Mound of the Dead." Only ruins of the city are left. Among them is one of the first pools ever built. It is the Great Bath. Scientists say the Great Bath is around 2,300 years old.

The ruins of Mohenjo-daro are in the northwestern part of Pakistan.

The Great Bath wasn't made for swimming. Instead, it was probably used for religious ceremonies. Builders used baked bricks to build the floor and walls. They covered the bricks with a natural tar. This tar kept the water in.

The Great Bath was about 13 yards long and 8 yards wide. What was the bath's perimeter? What was its area?

 Did You Know?

Gaius Maecenas built the first heated swimming pool. Maecenas, a wealthy man, served the Roman emperor Augustus. He also supported great poets of his time.

Greeks and Romans were probably doing the backstroke thousands of years ago. But modern swimming pools did not become popular until the 1800s. That's when English workers built six indoor pools in London, England.

The oldest concrete swimming pool in the United States is in Austin, Texas. At first, Deep Eddy swimming pool was a swimming hole. A large boulder sat in the Colorado River. Water rushed around it, forming an eddy. Swimmers swam in the "Deep Eddy."

Austin is home to Barton Springs Pool, too. It's 900 feet long and fed by spring waters.

In 1915, A.J. Eilers, Sr., bought the land near the hole. He built a concrete pool. He built cabins and camping sites. There were also shows like Lorena's Diving Horse to attract tourists. Lorena and her horse would climb a ramp to a diving platform. The platform was 50 feet high. Then, they would dive into the deep end of the pool. The Deep Eddy swimming pool is a rectangle about 33 yards wide and 68 yards long. What is the pool's perimeter? What is its area?

Time to Reflect

The Reflecting Pool is in Washington, D.C. The pool stretches between the Washington Monument and the Lincoln Memorial. Stand at one memorial, and look to the opposite end. You'll see a reflection of the other memorial in the water.

Henry Bacon designed the Lincoln Memorial in honor of Abraham Lincoln. Lincoln was the 16th president of the United States. He led the country during the Civil War.

Thirty-six columns surround the memorial. Each column represents a state that existed when Lincoln was president. The memorial opened to the public in 1922.

The Reflecting Pool is a rectangle about 676 yards long and 56 yards wide. What is its perimeter and area?

Math at Work

Some engineers are builders. They build skyscrapers. They build dams, tunnels, and bridges. And they build monuments like the Washington Monument.

The Washington Monument stands more than 550 feet tall. Its walls are mostly marble. They are 15 feet thick at the bottom and 18 inches thick at the top. The monument weighs 81,120 tons.

For some time, the monument needed repair. So, in 1996, groups joined forces to raise money to get the job done.

Workers sealed, fixed, and cleaned the monument. They sealed 500 feet of cracks in the stone. They fixed 1,000 square feet of broken and patched stone. They cleaned 59,000 square feet of wall surface.

The deepest part of the Reflecting Pool is 18 inches. But the pool holds almost 7 million gallons of water.

Swimmers must overcome drag to swim faster.

Connecting to Science

What a Drag

When you bike, there is friction between the wheels on your bike and the road. When you swim, there is friction between your body and the water. This friction is called **drag**, and it can slow you down. It slows down Olympic swimmers, too. So, they wear special suits.

Some Olympic swimmers wear full-body suits. The suits look like shark's skin. They are covered in tiny fins. The suits are made of different fabrics in different places. There is rough fabric where water flows more quickly. There is smooth fabric where water flows more slowly.

All swimmers experience drag when they move through water. One company says that their high-tech suits can reduce drag for swimmers. But you don't need a high-tech suit to paddle around in a backyard pool. A regular bathing suit will do.

There's a lot of water in an Olympic pool. Some Olympic pools are 164 feet long, 82 feet wide, and 7 feet deep. What is the perimeter of this rectangular pool? What customary units of capacity would you use to measure liquid volume, or how much water the pool holds?

 What's the Word?

If you like swimming, you might want to read about Gertrude Ederle. Check out *America's Champion Swimmer* by David A. Adler. You will learn how Gertrude broke world records in her swim across the English Channel.

After so much reading, James knows a lot more about pools. He wants his pool to look new, not old. And he definitely wants something smaller than an Olympic pool.

Aunt Patty draws a model of an idea she had for James's pool. "Do you like this shape?" she asks. "It has a square attached to a rectangle."

"Yes, yes, I like it!" James answers.

"Good. Well, let's take some measurements. Then, you can tell me if you still like it."

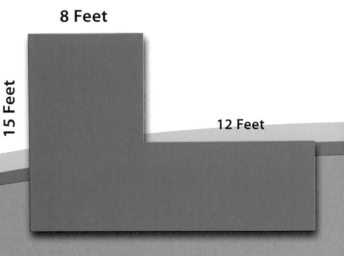

"Let's start with perimeter," Aunt Patty said.

Perimeter: Aunt Patty explains, "You can measure and add the lengths of the sides of the square and of the rectangle. Or, you can use the formula $2l + 2w$ for each shape."

"I think both ways work equally well. But, you have to figure out some of the measurements first. Use the measurements in the drawing to label every side of each rectangle. Then, use the formula you like better. What's the perimeter?"

"What about the area?" asked Aunt Patty. "Do you remember the two ways you can find area?"

Area: Uncle Ted jumps in with an explanation. "You can cover a pool's surface with unit squares to find its area, and then count the squares. But the unit squares will get wet, and measuring will be messy," he says.

"Or, you can use the formula $l \times w$ to find the area of each rectangle. Then, add the areas to find the total area. This may be better than using unit squares. What is the area of your pool?"

"Ahhh," says Aunt Patty. "Now it's time to fill the pool. What is the pool's **capacity?**"

Liquid Volume, or Capacity: You can use **customary units of capacity** or **metric units of capacity**. "In our work, your uncle and I use customary units. So, why don't we use those now? Which units shall we use?"

Aunt Patty brings out a chart James has seen before. "You could use any of the measures in this chart," she says. "But your pool is large. Which unit should we use to measure?"

teaspoon (tsp)	tablespoon (Tbsp) 1 Tbsp = 3 tbsp	cup (c) 1 c = 16 Tbsp
pint (pt) 1 pt = 2 c	quart (qt) 1 qt = 2 pt	gallon (g) 1 g = 4 qt

James makes some calculations and comes up with a final plan. The pool Patty and Ted designed is nifty. There will be plenty of room for James and some friends. Now he needs to start saving some money for some pool toys.

WHAT COMES NEXT?

Landscape designers design outdoor spaces. Sometimes, those spaces have pools and fountains. They almost always include plants. But not all plants will grow well everywhere.

Landscape designers have to know a lot about plants. They draw sketches of where plants will go in a garden. Before they sketch, they think about how much sunlight plants need. They also think about how systems will carry water to plants.

Design a garden you would like to have one day. Think about the colors and shapes you want in your garden. Then, go to the library, to a plant nursery, or online to learn about which plants grow best where you live. Pick some plants that will give you the color or shapes you want. Put them in your sketch.

Include a pond or fountain in your plans. Draw paths so that people can walk in your garden. Be your own landscape designer.

GLOSSARY

area: the measure of a space inside closed boundaries; Measures of area are in square units.

capacity: the amount of liquid a container holds; or liquid volume.

customary units of measurement: Customary units of measurement of capacity include teaspoons, cups, pints, quarts, and gallons.

depth: a measure of how deep something is.

drag: the friction between the water and a body when swimming.

legion: a group of Roman soldiers; a legion usually contained between 3,000 and 6,000 foot soldiers, plus soldiers on horseback.

length: a measure of how long something is.

liquid volume: how much liquid a container can hold; or capacity.

metric units of measurement: Metric units of measurement of capacity include milliliters (mL) and liters (L). One mL = 1/1000 of a liter.

mosaic: a picture or design made with small pieces of colored materials, such as glass or tile.

perimeter: the measure of the distance around a closed shape.

width: a measure of how wide something is.

FURTHER READING

FICTION
Mason Dixon: Pet Disasters, by Claudia Mills, Yearling, 2012
The Case of the Purple Pool, by Lewis B. Montgomery, Kane Press, 2011
NONFICTION
Perimeter, Area, and Volume, by David A. Adler, Holiday House, 2012
Sir Cumference and the Isle of Immeter, by Cindy Neuschwander, Charlesbridge Publishing, 2006

ADDITIONAL NOTES

The page references below provide answers to questions asked throughout the book. Questions whose answers will vary are not addressed.

Page 12: Perimeter: 50 inches

Page 13: Perimeter: 28 feet; Area: 40 square feet

Page 14: gallon

Page 16: Design 1: 44 square feet; Design 2: 96 square feet

Page 17: (12 + 3) + (12 + 3) + (8 + 3) + (8 + 3) = 52 feet

Page 20: Perimeter: 42 yards; Area: 104 square yards

Page 21: Perimeter: 202 yards; Area: 2,244 square yards

Page 22: Perimeter: 1,464 yards; Area: 37,856 square yards

Page 25: Perimeter: 492 feet; Unit of capacity: gallon

Page 27: Perimeter:
8 + 8 + 12 + 7 + 20 + 15 = 70 feet
Area of square: 64 square feet;
Area of rectangle: 140 square feet;
Area of entire pool:
64 + 140 = 204 square feet

Page 28: gallons

INDEX

CONTENT CONSULTANT

David T. Hughes

David is an experienced mathematics teacher, writer, presenter, and adviser. He serves as a consultant for the Partnership for Assessment of Readiness for College and Careers. David has also worked as the Senior Program Coordinator for the Charles A. Dana Center at The University of Texas at Austin and was an editor and contributor for the *Mathematics Standards in the Classroom* series.